Contents

Acknowledgements

The assistance of Andrew Larner, under contract to
CCTA from the University of East London is gratefully
acknowledged.

Geographic Information Systems:

A Buyer's Guide

Mike Gunston

CCTA

May 1993

London: HMSO

For further information on CCTA's GIS publications and services please contact:

Customer Services
Strategic Programmes Division
Gildengate House
Upper Green Lane
Norwich
NR3 1DW

For further information regarding this publication and other CCTA products please contact:

CCTA Library
Riverwalk House
157-161 Millbank
LONDON
SW1P 4RT

071-217-3331

1 Introduction and Overview

1.1 Purpose

Geographic Information Systems (GIS) can improve the effectiveness and efficiency of government business operations. Opportunities exist for government to increase the effectiveness of its business activity and reduce its costs through the use of GIS. Departments need to understand the issues, benefits and costs of introducing and using GIS. This GIS Buyer's Guide is intended to help readers gain the maximum benefit from their GIS investments by making good purchasing decisions without having to undertake extensive research.

1.2 Target audience

This guide assumes some knowledge of information systems and information technology. It has been written for those responsible for the purchase and use of GIS in government departments.

1.3 Structure

Much of this guide represents the consolidation and refinement of information that already exists elsewhere. It is assumed that this publication will be used as a reference guide and therefore some themes have been developed in different chapters from different perspectives.

A selection of questions to ask the supplier and considerations for the buyer to make are interspersed throughout the text. These are not exhaustive but it is hoped that they will nonetheless be helpful.

Chapter 2 describes the components of a GIS.

Chapter 3 provides an overview of the main issues surrounding GIS implementations.

Chapter 4 provides guidance on system specification and procurement.

Chapter 5 describes data types, sources and costs. It also covers the issue of copyright.

Chapter 6 includes an assessment of the market, its size and future trends.

Throughout this guide, the text necessarily includes some technical terms; these are described in the glossary.

1.4   Related publication

A supplement to the Buyer's Guide reviews major GIS suppliers, their background, products and services, financial details, strategy and details of GIS customers. This supplement is 'Commercial-in-Confidence' and is available to bona fide Government departments. A written request for copies should be made to the CCTA librarian.

2 GIS components

2.1   Introduction

There is a wide variety of software and hardware components associated with GIS, and organisations need to be selective about the composition of their GIS solution and the way in which GIS services are organised. The selection and arrangement of GIS components largely depends on the scale and nature of the business activities involved; there are, however, four GIS "architectures" which commonly occur, and which are appropriate for particular circumstances.

2.1.1   Personal GIS

Many business applications require only individual access to geographic information, and such applications can be supported by personal GIS solutions. Usually

such systems involve stand-alone PCs or workstations which often have direct access to inexpensive printing and/or plotting facilities such as a laser printer, inkjet printer, or small pen plotter.

It is common for personal systems to have digitising facilities attached, but many organisations prefer to use data conversion bureaux or existing CAD facilities for infrequent data entry needs. See Figure 2.1.

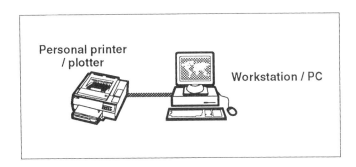

Figure 2.1: Personal GIS.

2.1.2 Departmental GIS

Applications which require more than one point of access within a user department are usually best served by "departmental GISs". Departmental GISs generally include a range of input, output, storage and processing components which are networked together and shared between a number of users. Organisations usually find it easier to justify expensive components when these serve a number of users; consequently departmental GISs tend to have better facilities than normally found on personal systems. Departmental GISs also tend to be integrated with other information systems serving the department. See Figure 2.2.

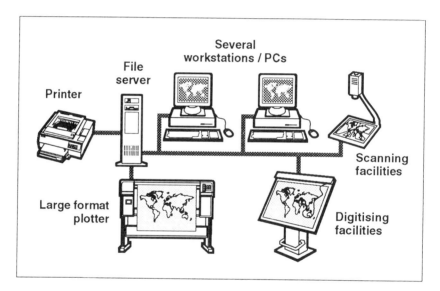

Figure 2.2: Departmental GIS.

2.1.3 Corporate GIS

In some organisations the exchange of geographic information between departments is an essential element of business practice. These organisations are best served by a "corporate" GIS architecture in which departments can obtain direct access to geographic information originating from other departments over a computer network. Corporate GISs also tend to be integrated with other corporate information systems.

Because user departments generally need immediate access to data capture and plotting facilities, most organisations tend to provide some facilities at departmental level, but more expensive components, such as scanners and large colour electrostatic plotters, are usually provided from a common GIS unit.

In some instances it is necessary to extend direct access to geographic information to remote offices. In such cases GIS services can be provided by methods ranging from routine data-file exchange to on-line access over Wide-Area-Networks. See Figure 2.3.

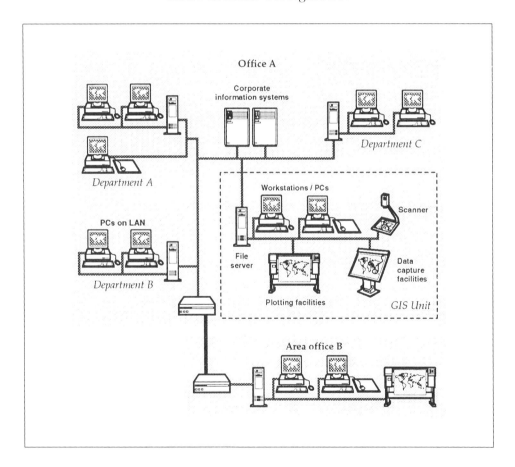

Figure 2.3: Corporate GIS.

2.1.4   Central GIS

Some organisations find it appropriate to provide GIS services from a central specialist unit. This is particularly appropriate where users do not require immediate access to information, have infrequent requirements, or where information processing requires particular expertise. With these scenarios, the technical architecture is similar to a departmental GIS, but the organisation recognises its multi-departmental role.

All the various GIS configurations outlined above utilize GIS components as illustrated in Figure 2.4, on the following page.

2.2   Hardware

2.2.1   Processing units

The hardware processing unit can be almost any type of computer platform, including mainframes, minicomputers, workstations, and, increasingly, high performance personal computers. The use of mainframe and minicomputers was considerable but the performance/cost ratio for workstations and personal computers has significantly improved during the last three years making them more popular for GIS. According to Dataquest [3], 57% of GIS world-wide were installed on workstations in 1991, 26% on host systems and servers, and 17% on PCs. The last is growing rapidly. The penetration in UK local authorities is 42% (workstations), 25% (mainframes) and 23% (microcomputers). Figure 2.4 shows the characteristics of computer processing units used for GIS applications.

Advances in technology have blurred the distinction between the terminology - mainframe, minicomputer and microcomputer. Performance levels for computers which were in the mainframe category are now well within the minicomputer range. The increase in the capabilities of desktop microcomputers has further obscured their distinction from minicomputers. The term workstation is usually used to describe a high performance device with a 32-bit processor, large main memory and high resolution graphics capability. These terms are therefore used to indicate general ranges of capabilities.

Component	Type	Attribute
Hardware	Processing unit	Mainframe Minicomputer Workstation PC
	Display	CRT Flat
	Storage device	Tape drive Disk drive Optical disk
	Data capture device	Digitiser Optical scanner
	Output device	Line printer Impact printer Ink-jet printer Laser printer Pen plotter Electrostatic plotter Ink-jet plotter
Software	Operating system	Proprietary Open
	Graphical user interface	Various
	DBMS	Flat file Relational
	Applications software	Many
	Other support programs	Many
Data	Graphic	Raster, vector
	Non graphic	Various
Services	Consultancy	Various
	Data capture/conversion	Various

Figure 2.4: GIS Components.

Mainframes

Some of the well known GIS can be run on various mainframe platforms (eg. Alper GIS on IBM, ICL and Bull, and ARC/INFO on IBM) although the trend is to port the software to minicomputers/workstations.

Minicomputers

In the minicomputer category, the Digital Equipment Corporation VAX series is frequently used for GIS software, although Data General and Hewlett-Packard processing systems are also popular.

Workstations

Workstations such as the Sun Sparcstation series, the Hewlett-Packard 9000 series, Intergraph's C400 series, IBM's RISC Systems 6000 series, ICL's DRS 6000 series and DEC VAXstations are used by GIS software vendors as integral parts of their systems. They mostly use the UNIX operating system and increasingly employ Reduced Instruction Set Computing (RISC) processors which require large main memories.

Processing type	Number of users	Word architecture	Main memory	Role	Cost range
Mainframe	More than 256	32-bit or greater	64MB+	Volume processing	£250K-£20M
Minicomputer	Up to 256 devices	32-bit or greater	8MB-256MB	File server	£50-250K
Workstation	Single user	32-bit	4MB-64MB	Stand-alone	£3-50K
PC	Single user	16-32 biit	1MB-16MB	Stand-alone or single user node on network	£1-8K

Figure 2.5: GIS Computer Processing Units.

PCs

GIS software uses microcomputers as:

- single user systems on a local area network running microcomputer GIS and CAD oriented mapping packages

- intelligent workstations attached to a host processing unit that perform some local processing but still require the host processing unit for most complex GIS functions

- non intelligent graphic workstations attached to a host processing unit to emulate graphic terminals for digitising, and graphic queries

- data capture stations that batch-load data to and from a host processing unit.

Micro-processors

Personal computers use micro-processors from Intel and Motorola. The standard PC operating system for Intel systems is MS-DOS, developed by IBM and Microsoft. The main alternative is the Apple Macintosh series of personal computers which use Motorola micro-processors and have their own proprietary operating system with auxiliary facilities for running DOS applications. The standard PC has significant use in the low-end single user market for mapping and database management operations.

Desktop personal computers based on 32 bit processors are becoming more popular as a low cost platform for GIS development. Together with the multi-user UNIX operating system, this platform can support sophisticated software, high resolution graphics, large memory and mass storage.

Distributed systems

There has been a shift from central processing to distributed systems where some tasks are distributed from a main host and handled by separate processors. There is also the ability to distribute the data. The growth of high speed networks has increased the capability for remote data communications and hastened the development of techniques and standards for linking different types of computers.

Displays

The screen refresh speed, flicker rate, size, resolution and quality of colour Cathode Ray Tube (CRT) displays has been improving steadily. Displays of one megapixel (1000x1000 pixel or screen dot image) are common in high-end workstations and are becoming more prevalent on PCs and graphics terminals. The availability of graphic processors enables the larger displays to maintain speed. Low cost memory is increasing the colour capability of lower range displays, including 24 bit colour on PCs. High-end displays can simultaneously represent 16.7 million colours (ie. each pixel is represented by 24 bits of memory). The technology of flat screens has advanced rapidly but these screens do not approach good CRTs in quality, speed, size or colour. Colour Liquid Crystal Display screens will become more widely used as prices fall.

Questions to ask the supplier

Ask about the proposed processing units;

- How was the required power estimated?
- How were the number and types of unit chosen?
- What role does each of the units perform?
- Were any assumptions made about the use of existing equipment?

Ask about the provision of communications facilities;

- How were data transfer capacities established?
- Has the supplier previous experience of distributed systems?
- Who will install, configure and maintain communications equipment?
- Are there third-party costs such as line rental?

2.2.2   **Peripherals**

In addition to the standard storage and processing devices, specialist peripherals are required for data input and output.

Storage devices

Tape drives - Magnetic tape cartridges and reel to reel drives are used in geographic information systems for archiving and transporting large volumes of data and software. Tape drives provide an effective and low cost method of storing data off-line (a tape must be mounted physically before data can be read). Digital Audio Tape (DAT) and video cartridges are becoming more popular because of their capacities and compactness.

Disk drives are direct access storage devices which allow users to store and retrieve data directly without loading from another source. They use magnetic or optical technology to store information and are connected to the processing unit through high speed channels so that data can be retrieved directly by system users.

Optical disks are read-only, write-once or re-writable. Recording densities are significantly higher than with magnetic media and the medium is less prone to damage. Re-writable optical cartridges can be removed from the drives so that the data can be transported to other locations. This capability gives the re-writable optical disk considerable versatility. An optical disk jukebox is a high performance robotic system which provides large storage capacity with a high speed retrieval system for fast data access.

Data capture devices

Data capture includes the process of image digitisation plus all subsequent processing, conversion and data communications that have to occur before the data are ready to use in their computer format.

The **digitiser** is a peripheral device for converting graphic information from an analogue to a digital form. Maps are re-traced using a digitising table and cursor or an analytical stereoplotter (for photogrammetric digitising).

Optical scanners digitally encode information from hard copy maps and documents. The scanner senses variations in reflected light from the surface of the document. The digital file that results from the scanning process is in raster (grid) format and does not generally differentiate between different types of features appearing on the original hard copy map. The raster image is a non-intelligent picture. Most scanning systems have software which converts the raster image to a vector format. (Refer to Chapter 3, Section 3.6).

Output devices

Dot-matrix **line printers** are used for high volume printing of reports, tables and other text documents. They are popular because of their high speed and relatively low cost. Printing speeds vary from 300 to 2,000 lines per minute. Printers in the range of 600 to 1,200 lines per minute are typical in geographic information systems that require volume text output.

Impact printers use a dot matrix to print graphics. The raster image is produced as a pattern of dots generated by hammers striking a print head over a ribbon. Multiple colour ribbons can be used to generate a small range of colours. Dot matrix printer resolution is limited by the density of pins on the print head; most offer a graphic resolution of 70 to 150 dots per inch. Ribbons have a limitation for generating high quality colour products. Overall clarity tends to be inconsistent and the printers are noisy, but costs are relatively low.

There are a number of colour **ink-jet printers** which are relatively cheap and can produce a maximum resolution of 300 dots per inch. However, printing speed is slow and the quality of ink-jet output is disappointing compared with professional colour printing.

Page printers, such as **laser printers** are used to produce quality hard-copy text and graphics. They are being used more extensively in GIS because of their versatility, low noise and decreasing price. Several manufacturers offer colour models and their prices are falling rapidly. High volume laser printers deliver speeds of up to 30 pages per minute with a resolution of 300 dots per inch. Printers with a resolution of 400 and 600 dots per inch are starting to appear. Lexmark and QMS, among others, have introduced hardware with 600 dots per inch printing capability. Hewlett-Packard's LaserJet 4 is based on the 600 dots per inch LBP-EX Canon engine. To cope with higher resolution, there is a move towards more capable printers which incorporate a built-in RISC processor.

Pen plotters can plot vector format data using multiple line weights and colours with a high resolution. In GIS, drum-feed plotters are the most popular because they can strike multiple plots continuously using roll media, the plot length is flexible and costs are competitive. Plotters range from small desktop units to large format devices. Generally, pen plotters are used in GIS for large format products (24 to 44 inches width). Plotters can house multiple pens and can accept different input media. Host software produces the map data in vector form with a series of plotting commands which are part of a library that specific vendors use as a convention on their plotters. Maximum plot speeds exceed 30 inches per second. There are 'intelligent' plotters which can arrange vectors in the plot file accepted from the host system, to make plotting more efficient (Refer to Chapter 4, Section 4.5.5).

Electrostatic plotters connected to a host processor read graphic files and produce hard copy plots. They are used in the high volume production of plots generated by GIS (they are much faster than pen plotters). The plotter reads a plot file in raster form in which the image is composed of a matrix of dots. An image is then produced by the electrostatic plotter using the same technique as a photocopy machine. The speed can be significantly slower if vector information from the GIS is converted to a raster format for the plotter to read it. Plotters tend to be equipped with local controllers and rasterising units so vector information can be sent from the host to the plotter and rasterised there. Unless specially compressed, raster files require much more storage space than their vector counterparts, so plotter rasterised units include storage buffers. Plotters can generally print at 400 dot per inch resolution on opaque paper and clear and matt finish polyester film (colour and monochrome output).

Ink jet plotters force coloured inks through small jets on to the print media. Resolution ranges from 120-300 dots per inch. These plotters have become more popular for GIS due to significant improvements in speed, resolution and reliability. Plotter prices generally have fallen substantially.

Considerations

Consider alternative ways in which data capture could be undertaken, such as:

- Hiring staff and equipment
- Using a data conversion bureaux

When planning input and output facilities consider:

- Who needs to have direct access to facilities?
- Who needs to have shared access to facilities?
- How often will the facilities be used?
- What quality of input/output is required?
- What skills will be required to use the facilities?

Questions to ask the supplier

Establish on what basis have data storage and transfer devices been proposed;

- What is the role of each device?
- How will data be loaded into the system?
- How has storage capacity been estimated?

What provision has been made for data conversion facilities;

- What is the speed of use/throughput of the proposed facilities?
- What is the size and nature of the data conversion programme?
- How are the facilities to be made available within the organisation?
- What training is required to operate the facilities?

Establish on what basis have output facilities been proposed;

- what are the running costs of the devices?
- which facilities are for departmental, central, or personal use?
- which facilities require operator supervision?
- what is the maintenance regime and life-expectancy of the devices?

2.3   Software

A GIS is supported by three types of computer software: the operating system, the applications software and auxiliary support programs. Users have become further removed from the actual operation of the hardware as the software has become more user friendly.

2.3.1 Operating system

The main functions of the operating system include memory management, system access and accounts, communications control, command processing, file and data management. A GIS is no different from any other information system in that it requires all the normal features of an operating system.

Until recently, most operating systems software was proprietary, designed to run only on a particular family of computers offered by a single vendor. The operating system was acquired as part of the overall hardware package and was installed and activated by the computer supplier. The use of proprietary operating systems is still prevalent in the mainframe and some minicomputer ranges. Operating systems predominant in geographic information systems include VMS (DEC VAX minicomputers) and MVS (IBM mainframe).

GIS software is becoming available on a wide range of hardware which supports the UNIX operating system. UNIX was originally developed for scientific applications by the University of California and AT & T Bell Laboratories. Most hardware manufacturers offer UNIX with their mid-range systems and workstations because of the increasing portability of applications. The concept of portability means that any application package can be loaded on another processing unit regardless of make or model and can operate without program modification. This concept has not been fully realised. Most users are confined to particular hardware environments unless they perform program modification. Moreover, there are no recognised standards specifically for GIS applications.

Questions to ask the supplier

Concerning the operating system:

- Do existing operating systems need to be upgraded?
- What are the facilities for communicating with other systems?
- What are the maintenance costs?

2.3.2 Applications software

GIS applications software packages consist of multiple programs that are integrated to supply particular capabilities for mapping, management and analysis of geographic data. The package usually has core utilities for basic mapping and data management functions which are integrated with separate modules for specific GIS applications. Software packages designed principally for mapping may emphasise graphics processing and have limited database functions and geographic analysis utilities. Packages designed to be full geographic information systems usually include strong graphic and non graphic data management features. Special application modules may include functions such as network tracing and terrain (or elevation) analysis.

Graphics processing capabilities include functions that allow the user to enter or edit map features and to generate screen displays or hard copy maps. Functions traditionally associated with CAD systems have been incorporated into GIS map entry and edit packages.

Graphic entry capabilities in GIS allow users to input map features and store these as co-ordinates based on a plane reference grid. Text annotation and feature identifiers can also be entered to define a map feature uniquely and provide a basis for associating the feature with tabular attributes stored in a database. Core package graphic entry capabilities usually include interactive digitising and special-feature entry.

Annotation entry - GIS packages create annotation, text to be displayed on maps, by position and entry during the map digitising and editing processes. Alternatively, the non-graphic database can be accessed and data elements can be used within it as annotation for specific map features. Some systems will allow the customisation of a map display. More advanced systems will allow parameters for displaying annotation to be swapped around.

Graphic editing - GIS application packages have a variety of feature delete and modification functions and some have edit programs which are executed in batch mode on a digitised map file. Some graphics software can also generalise or smooth lines stored in vector form.

Graphic display - The display has software capabilities that enable an operator to control the appearance and format of the screen or plot by the setting of parameters (eg. setting line style or map scale). Some GIS have software capabilities that enable the operator to create or design a database through a graphical user interface.

Database management - GIS application packages usually include capabilities to store and retrieve non-graphic attribute data associated with map features. GIS systems store non-graphic attributes and link them with their associated map features to support map display and analysis. These packages use proprietary or third party database management software to manage the non-graphic database. Non-graphic database software used in GIS uses a data definition language (DDL) which allows a user to describe the characteristics of files that contain non-graphic attributes. Database management systems (DBMS) offer some degree of interactive non-graphic data entry. Data can be extracted from the database according to user defined criteria using a data manipulation query language.

A GIS can be used to query and analyse both graphic and non-graphic data together and some database management systems include features for complex report design using fourth generation languages. There are database independent development platforms which permit applications and users to interact with relational database management systems (RDBMS). Most GIS suppliers support the popular database management systems such as Informix, Ingres, Oracle, DB2 and dBase. By incorporating ISO standard SQL (ISO 9075) and standard networking protocols, database development platforms can eliminate the need for maintaining separate interfaces for each RDBMS that an application uses.

Special Utilities - GIS packages have special utilities for performing routine mapping and geographic analysis. These utilities can be used alone in simple applications or combined with others to build more complex applications. Examples are map co-ordinate and geometric transformation utility programs and basic cartographic operation utility programs for edge matching and map merging. In some packages, programs are included as part of the core software to perform special map production by graphic overlay thematic mapping (superimposing a set of graphical data on a base map for depicting specific information) and address matching.

Some packages explicitly store map features as objects, allowing a more in-depth analysis of the area covered. (Refer to topology, Chapter 3, Section 3.3.1).

GIS software packages should include utilities for map analysis eg distance and area measurement, radius search, and appropriate statistical tools. Other utility programs include functions for raster/vector conversion and overlay, and data transfer.

Availability and choice - There has been a rapid increase in the number of vendors entering the GIS software market and the functionality offered by GIS packages. Applications software is available for hardware platforms ranging from mainframe systems to microcomputers. The selection of GIS software can be complex because the market is dynamic and products vary in approach, design and customisation potential. Choice should be based on an evaluation of product capabilities and is considered in Chapter 4, Section 4.5.

Considerations

Refer to the LGMB publication GIS: Functional Specification for technical questions regarding generic GIS facilities.

Make sure the system selection team has the necessary experience to make decisions on financial, management, user and technical issues.

Questions to ask the supplier

Ask for an explaination of GIS facilities:

- What are the facilities for drawing/editing map features?
- What gazetteer facilities are there?
- Can textual attributes be accessed?

What application development facilities are there for:

- Defining screen and report layouts?
- Defining menus, macros, attributes?
- Producing a library of graphic symbols and line styles?
- Automating processes?

Concerning software support:

- What is the cost of maintenance/user licenses?
- What are the costs and frequency of upgrades?
- Are there telephone helplines?
- What are the procedures for error reporting and correction?
- What application development services are there?

2.3.3 Support programs

Most IT systems have program modules that are accessed by the operating system or application packages to perform routine support functions. Major utilities and support software used in GIS include language compilers, device drivers, disk backup utilities and communications software.

Device drivers are programs which provide a communication interface between the operating system and the application software to support output on a peripheral device. In Geographic Information Systems, device drivers are used to support plotters and other graphic output devices, eg. device drivers interpret specific plot generation commands.

User interface - Most GIS software packages employ some level of screen menu driven graphical user interface (GUI). Users point to areas on the screen using a mouse or digitising cursor, to issue software operation commands.

At a higher level, the GUI is used to invoke housekeeping tasks generally associated with operating systems, such as file management and applications control. The popular iconic GUIs deployed on personal computers are the proprietary Microsoft Windows, and the Apple Macintosh interface. These provide multiple application contexts concurrently on a single display.

The X Window system process running on a workstation, provides a display and windowing service to application processes running on the workstation or elsewhere. X Window provides the mechanism for drawing windows, it does not define a particular user interface.

Questions to ask the supplier

Concerning systems administration, what facilities will be provided for:

- Data backup and recovery?
- Printer and plotter queue management?
- User administration and security?
- Data import/export?
- Scanning?
- Raster-to-vector conversion?
- Base map management and editing?

2.4   Data

Data is the essential raw material for a GIS. It can be varied and complex with digitised map features and non-graphic data representing geographic location characteristics. Large volumes of data can be expensive to collect and store, accounting for the greater proportion of the total GIS implementation cost. The availability and volume of data has increased as a result of national mapping programmes, the creation of global environmental databases and the use of remote sensing satellites. A GIS may use data residing in administrative or business areas as a source for spatially referenced digital data.

As discussed in Chapter 3, Sections 3.1 and 3.2, graphic and non-graphic data have distinct characteristics and there are different techniques for their management. A GIS usually has a data base management system that will manage at least the non-graphic data. A geographic database may consist of map features and associated non-graphic attribute elements. Database development often requires the conversion of analogue maps to a digitised form or the compilation of digital maps from aerial photographs or field data.

In most current systems, there is some distribution of processing power and sometimes of data. The potential need to exchange data between systems to support GIS applications, is extensive. Methods used to transfer data between computer systems vary in complexity.

The main data issues common to GIS applications are quality, accuracy, volume, complexity, availability, capture and standards. These are described in Chapter 3.

Considerations

As far as possible, establish the availability and suitability of external geographic datasets as part of your Feasibility Study. The Tradeable Information Initiative Metadata disk, available from AGI, catalogues many government datasets.

> **Questions to ask the supplier**
>
> Concerning the nature of the information:
>
> - How accurate is it?
> - What is its source and scale?
> - Does it sufficiently cover your organisation's
>   stated requirements?
> - How old is it, what is its frequency of update?
> - What are its transfer format and media?
>
> Concerning the the costs of the information:
>
> - What is the purchase price?
> - What are the maintenance costs?
> - Whose is the copyright and are user licenses required?
> - What implications are there for derived data?
> - Are the charges stable?

2.5 Skills

Whilst some definitions of GIS may not include skills as a component, these are essential for the successful purchase and implementation of a system. A GIS requires skilled people for its successful design, implementation, operation and support. A GIS is not merely a database with a spatial dimension. The introduction of the extra dimension creates problems of visualisation, literacy and interpretation, which are by no means trivial. Buyers should be aware of the need for education into a different way of thinking when implementing GIS. Most GIS need modification as their development tends to iterate. Consultancy services are often used for the design, development, implementation and post-implementation changes of GIS. (Refer to Chapter 4, Section 4.4). Suppliers tend to identify these services as the main area of differentiation between their system(s) and the competition.

There should be a project manager responsible for managing the acquisition of the system, development of the data base, assignment and training of staff, application program development and overall system preparation. Skilled in-house staff are usually used for day to day management of all system components and operations, led by a system manager. A database administrator would control the resources, quality and maintenance of the database, and may use bureau skills and temporary staff for data capture and entry. Alternatively, data capture can be outsourced. Purchasing considerations are discussed in Chapter 4, Sections 4.3, 4.4 and 4.5.

> **Questions to ask the supplier**
>
> Concerning training:
>
> - What types of training are available?
> - What previous experience is assumed?
> - What does training cost?
> - Where does training take place?
> - What training facilities are provided?
>
> Concerning documentation:
>
> - What is available for.GIS users?
> - What is available for systems administrators/operators?
> - What is available systems developers?
> - What other documentation is available?

2.6 References

[1] Antenucci J., Brown K., Croswell P., Kevany M., Archer H.

'Geographic Information Systems - A guide to the technology' 1991
Van Nostrand Reinhold
ISBN 0-442-00756-6

[2] Campbell H. and Masser I.

'The Impact of GIS on Local Government in Great Britain' 1991
Department of Town and Regional Planning (University of Sheffield)

[3] Dataquest Market Statistics

CAD/CAM/CAE and GIS Applications (May 1992)
Dataquest Inc, San Jose CA 95131-2398

[4] Maguire D.

'An overview and definition of GIS'
'Geographic Information Systems' 1991 Volume 1 P9-20
Longman Scientific and Technical ISBN 0-582-05661-6

[5] UNICOM

Greener A., Hart A., Pearson E., Tulip A.
'Geographic Information Systems'
Draft Information Technology Report 1992

[6] Local Government Management Board

'Local Government Geographic Information Systems
Product Software Function Requirements Specification'
1991
ISBN 0-7488-9876X

LGMB
41 Belgrave Square, London, SW1X 8NZ.
071-235-6081

3 Geographic information issues

3.1 Introduction

Building a GIS database is the most expensive, time consuming and problematic aspect of implementing a Geographic Information System (GIS). The issues surrounding GIS databases are very different from those of conventional databases. An understanding of the basic elements of a GIS database and its structure is essential in order to ensure effective and efficient implementation. This chapter provides a simple overview and should assist business IT managers with understanding much of the terminology surrounding GIS implementations.

A GIS database attempts to describe, at an appropriate level of detail, a specific view of a collection of objects (points, lines and areas) in the real world. The level of detail is largely determined by the use to which the data will be put, but may also be affected by limitations imposed by technology. The database may contain two distinct types of data: graphic data and non-graphic data.

Graphic data are a digital representation of the spatial characteristics of an object, such as length, position, area, volume or shape, or of the spatial relationships of an object relative to other objects in the database, eg. inside, outside, left or right. The non-graphic data represent the non-spatial characteristics of an object, such as name or function; these data are usually stored as alphanumeric text and may already exist in an organisation, held in conventional database format. Post codes are a form of geographic data with which most people are familiar; in this instance the graphic data might describe the boundaries of the area covered by an individual code, while the non-graphic data may contain, for example, census information for that code. The post code may itself be included in the non-graphic data (i.e. it is a name attached to the defined area) or it may be the link between the graphic and the non-graphic data.

Graphic data and non-graphic data are normally stored and manipulated separately for efficient processing. The problems associated with the collection and maintenance of non-graphic data are no different to those for conventional IT systems and are not therefore covered in this publication. The rest of this chapter concentrates mainly on the issues surrounding graphic data.

3.2   Graphic data

The geographic data held in a GIS database represent only one view of a real world situation; there are many ways in which the real world can be represented, depending on the requirement. The simplest example of different representations is one of scale; a town may appear as a single dot on a small scale map, but on a larger scale map the same town may be displayed as individual buildings and streets. In another example, the exact distance between points on a map may be irrelevant, nor is it always necessary to have accurate co-ordinate information where only the relationship between the objects is crucial. For example, while the London Underground map clearly displays the relationship between the individual stations on the network, it is not exact in terms of distance or precise geographical locations.

Graphic data are used to generate a map or cartographic picture on a display device, which may be a VDU, a printer or a plotter. The basic constructs of graphic data are elements that are referred to as points, lines and areas.

An object represented as a point may be stored in a GIS database as a pair of co-ordinates that identify a location in the real world. A line is a set of points which may represent linear features such as roads or rivers, but which may also refer to a non-physical boundary, such as a county boundary. An area is a collection of lines that close to form a discrete unit.

All graphic elements have so far been described in simplistic two dimensional terms. Points can exist in three dimensional space and lines may also exist in three dimensions, tracing convoluted paths through volumes. Some GIS software can now handle the display of 3D landscape images; this is sometimes referred to as Digital Terrain Modelling (DTM), or because it is a 2D representation of three dimensions it is also referred to as two and a half dimensions (2.5D). A fourth dimension can also be introduced to a GIS system in the form of time, either as life histories or as version management. The need to maintain different versions of a GIS database is covered further in Section 3.7.

Each object in a GIS can be associated with a number of attributes (non graphic data) which describe its characteristics at an appropriate level of detail. Attributes are linked to the graphic elements through a unique identifier that is stored in both graphic and non-graphic records. Attribute data may be managed by the GIS directly, or it may be acquired as needed by a related, but separate, database management system. It may also be analysed or queried by conventional methods that do not require reference to location.

Text, or label information, is commonly placed on printed maps; the text may refer to street or river names but may also be numeric values to indicate elevations, dimensions or addresses. Similarly, this information will need to be displayed or printed from a GIS. Sometimes known as annotation, this information is stored as text but will also require a combination of orientation rules, co-ordinates and size that dictate how and where it is to be displayed or printed. A variety of fonts for both screen display and printing is a normal requirement of a GIS.

Graphic symbols are also commonly used on printed maps to represent for example, parking areas, lighthouses or camp sites. Similarly in a GIS, graphic symbols can be attached to points, lines and areas for display purposes. Most GIS systems will include a symbols library together with facilities for creating and storing user defined symbols. Alternatively, additional information can be displayed by means of varying the display attributes of objects; for example lines can be represented in a variety of ways, such as solid, dashed and dotted lines of varying thickness, each having its own meaning.

Queries involving a GIS database may result in graphic display but may also result in a simple table or list of data elements. Most GIS implementations will combine both graphic and non-graphic displays. It is essential to define at an early stage the spatial elements, attributes and relationships that are appropriate to the requirement, and to decide what types of objects and data, described simply above, are to be handled. It will then be necessary to decide how the data should be structured for physical storage.

Questions to ask the supplier

How will the geographic information be presented:

- On the screen?
- On plotted output?

How will the geographic information be represented and processed within the system?

3.3 Data structure

In a GIS implementation, graphic data usually represent map images generated by computer from spatial data held in a GIS database. These data can be stored in one of two ways: vector format or raster format.

3.3.1 Vector

Vector data are represented by co-ordinates or strings of co-ordinates representing the position of objects (points, lines and areas).

In its simplest form, objects are represented by strings of co-ordinates; a point, for example, will be held as a single set of co-ordinates, a line as a set of co-ordinate pairs denoting its start, intermediate and end points, and an area as a string of co-ordinates denoting the boundary lines. No object is explicitly connected to any other object in the dataset and the data are not physically stored in any logical sequence. Thus a road intersection, for example, may be visually obvious when displayed, but unless that intersection is specifically recorded as an object, the retrieval of the appropriate information from the database can be time consuming. The subsequent calculation of the precise location of the intersection is not simple. Similarly the identification of an object as being within a defined area presents a difficult calculation. This loose and unstructured collection of graphic objects, known as a spaghetti data structure, is ideal for cartographic applications where the visual image, displayed or printed, is of primary importance. However, for those applications where some form of geographical search or analysis is required, a more complex data structure is required.

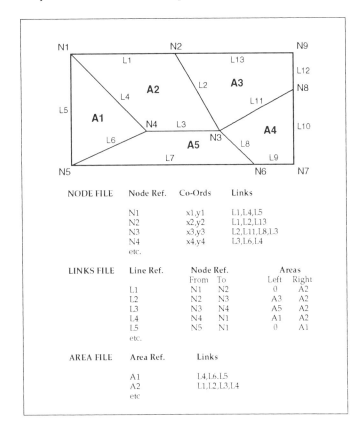

Figure 3.1: Topological data structure.

A topological data structure is concerned with establishing the location of objects with respect to each other, by defining connectivity, adjacency and containment. In this data structure, end points and intersections of lines are recorded as nodes, the lines themselves are identified as links between the nodes and the enclosed areas defined by a chain of lines are recorded as polygons. Links and nodes structures also allow spatial relationships (such as left, right, inside and outside) to be explicitly defined. Although this type of structure is more complex than spaghetti data, geographic analysis becomes much simpler. Updating the data also becomes simpler as co-ordinate information for a line is shared by adjacent areas, and node information is shared by connecting lines. A simple example of how a topological data structure might be applied is shown in Figure 3.1.

3.3.2   Raster

In a raster data format, graphic images are stored as a two-dimensional matrix of uniform grid cells (pixels). Each pixel is assigned a value which indicates the display attributes of that pixel. In simplest terms this value may be value 0 or 1, denoting white or black for display purposes, but may be assigned any value denoting colour or greyscale variations. The quality and clarity of the raster image depends on the number of pixels (or resolution) which in turn depends on the size of the grid cell. The larger the pixel size, the less precise the information; the smaller the pixel size, the larger the volume of data to be stored for any given area to be covered.

The raster data format is generally very simple and has advantages in some types of analysis; for example, overlaying two sets of raster data is simply a matter of adding or subtracting the values assigned to individual pixels. However, it does have a major disadvantage; raster data can be very large. While the consequent disc storage overheads are obvious, the processing overheads are sometimes overlooked. The retrieval time for a complete raster image can be lengthy, while the need to hold the image in memory while it remains in use, can significantly affect the speed of calculations or searches.

In order to overcome these problems a number of techniques have been developed to reduce the size of raster data. The two most common methods used are run-length encoding and quadtrees.

Run length encoding uses the fact that objects frequently extend over areas larger than a single pixel. Instead of recording the values of each individual pixel, this method groups the rows of a raster matrix into blocks with an identical value. For example, the values attached to a row of pixels representing a black and white image may be 000000111100000; using run-length encoding this would be recorded as 604150 (six zeroes, four ones and five zeroes).

In the more commonly used quadtree technique, the map is broken down into a hierarchical structure of squares with a common value. A simple example is shown in Figure 3.2. The map is first divided into four quadrants; if one quadrant has all the same value (as for the top left hand quadrant in Figure 3.2), that value is coded once only for the whole quarter. The remaining quadrants are further subdivided into quarters, until all quadrants so created have a common value.

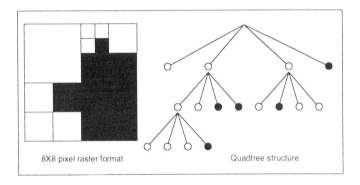

8X8 pixel raster format   Quadtree structure

Figure 3.2: Quadtree data structure.

The decision as to which data structure is appropriate is largely dependent on the requirement, although certain types of data will be captured in particular format eg. satellite images as raster data. Figure 3.3 shows a simple and very generalised summary of the advantages and disadvantages of vector and raster data.

	Raster	Vector
Data capture	Fast	Slow
Data volumes	Large	Small
Graphics	Medium	Good
Data structure	Simple	Complex
Geometrical accuracy	Low	High
Linear network analysis	Poor	Good
Area/polygon analysis	Good	Poor
Combining data layers	Good	Poor
Generalisation	Simple	Complex

Figure 3.3: Advantages and disadvantages of raster and vector data.

Considerations

- Scanned base maps are very much cheaper than vector base maps. Establish which features your organisation really needs to have in vector form, it is usually only a few features for each application. Consider whether these features should be purchased as a ready-made product (such as an OS digital map), or captured as part of your organisation's data conversion programme. Do not assume you have to buy vector base map data.

- Before purchasing the data ask for a trial dataset to establish whether there are any problems in using it.

Questions to ask the supplier

Ask the system supplier to confirm the suitability of the proposed data sources.

3.3.3 Layers

Whatever the format used, a detailed map of even a small area can be immensely large and complex if all features are recorded in a single data structure, making analysis and maintenance of the data very difficult. To overcome this problem the graphic component of a GIS database is often described as a series of layers, allowing individual or multiple layers to be more simply selected for display or analysis as required. Each layer is a set of homogeneous features that relates to the other database layers through a common co-ordinate system. The sequence of layers usually begins with a reference grid, on top of which other layers will describe application specific information, for example, administrative boundaries or utility features. A change in the base layer may also require associated changes to other layers.

In a topological data structure, co-ordinates are stored once only in a single layer. Each subsequent layer would be used to hold logically connected information. For example, one layer might hold details of house boundaries while another might hold details of soil types. While these layers are linked through a common co-ordinate layer, there is no logical connection between the distribution of these two layers, ie. property boundaries do not precisely follow soil boundaries. However property boundaries are relevant to many types of logically connected information, eg. land ownership or land use, and will therefore be stored within a property boundaries layer. Any change to the topological co-ordinate information relating to the property boundaries would not affect the soils layer and may therefore be totally transparent to soils applications.

3.4   Co-ordinate systems

Since maps were first produced, one of the main problems faced by cartographers was how to project the earth's surface on to a two-dimensional flat surface. In all cases true distances and directions may be compressed or stretched in order to fit the flat surface. A number of different methods are in common use by cartographers; the problems for GIS users arise when data are derived from a number of sources which may not all use the same projection method. For example, even an aerial photograph will not be an accurate representation of the earth's surface, in terms of exact distance; if this information is combined with measurements taken from a land survey, some discrepancies will inevitably occur. Over small areas, even the size of a small town, it is possible to treat the earth's surface as flat. Even a relatively small country like Britain may be treated as a flat surface, except where a very large scale, detailed map is required. If further reading on map projections is required, see references [2], [5] and [6].

Considerable reference has already been made to co-ordinate systems, which are used to describe the precise location of an object. GIS data can be represented using any one of a number of different co-ordinate systems. The most commonly used are geographical co-ordinates of latitude and longitude, or simple plane co-ordinates (x,y) measured from a fixed point. These co-ordinate systems are further explained in Figures 3.4 and 3.5.

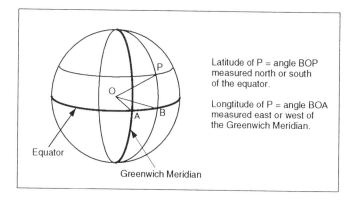

Figure 3.4: Geographical co-ordinates.

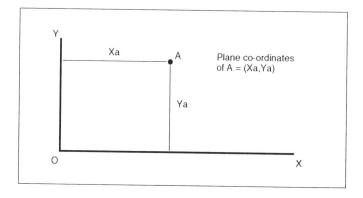

Figure 3.5: Plane co-ordinates.

In order to perform any form of geographic analysis, the information needs to be in the same co-ordinate system. Data using different co-ordinate systems can be converted using transformation facilities that should be included in most GIS software packages.

| 3.5 | Data capture |

Data are a crucial aspect in the successful implementation of a GIS, since a substantial portion of the data needs to be available before production operations can begin. The non-graphic data may already exist within an organisation. However few organisations will already have graphic data, which will therefore need to be purchased. Purchased data are unlikely to meet the precise needs of any organisation, therefore some form of data capture exercise is inevitable in order to obtain the specific information required.

The cost of the data capture exercise should not be underestimated. Significant costs are associated not only with the purchase of the necessary map data, but also with the coding of user or application specific information. It is common for the cost of data collection to exceed hardware and software costs by a factor of two. Some estimates indicate that up to 80% of the total project costs are associated with data collection. The costs will vary widely and will depend on the scale and detail of the information required, and not least on the pricing policy of the organisation providing the base data.

Geographical data are expensive to collect because large volumes are normally required to solve substantial geographical problems. Even for relatively small systems, the data capture exercise can be prolonged, with the major benefits only being realised after some years. **There should be no unrealistic expectations of early benefits of a GIS.**

Most of the contents of geographic databases continue to be derived from paper maps. Current technology offers two main methods of creating data from paper maps: digitising which produces vector data, and scanning which produces raster data. Digitising requires a human operator to position a map on a table and move a cursor over it, thus capturing the position of points and building up a digital representation of point, line and area features. This process is error prone, tedious, time consuming and expensive. Scanning allows a much quicker and less costly method of capturing the entire contents of a document. However, most systems require some form of vector data. Interpreting features from the scanned image may be error prone and requires human intervention.

Maps were not designed for the purposes of geographic databases and are not always the most appropriate source of data. Advances in technology now offer a variety of alternative sources for geographic data. Data can be captured directly through remote sensing, aerial photography or ground survey. The development of Global Positioning Systems (GPS) also provides a relatively new technology for determining a position on the earth's surface to a reasonable degree of accuracy, although highly accurate data are not available in real time.

Increasingly there is a trend towards scanning raster images and using line-following techniques to convert to vector format. This is further discussed in Section 3.6 below.

While combining data from some or all of the above sources is possible, there will be differences in scale, projection and in the co-ordinate system used. The data will then need to be transformed and edited to ensure consistency.

Considerations

It is often useful to test the services of data conversion bureaux by asking them to process a sample area first. The AGI yearbook contains a list of companies which provide such services.

> **Questions to ask the data conversion bureaux**
>
> Specifically:
>
> Which techniques will be used for converting
> the information:
>
> - Manual digitising?
> - Raster-to-vector conversion?
>
> Generally:
>
> - What quality checks will be made before the data
>   is issued?
> - In what form will the data be delivered?
> - Under what terms can errors be corrected?
> - How many source records are required and over what
>   period of time?

3.6 Restructuring and editing

Once the geographic data has been entered into the system, procedures for the detection and correction of errors will need to be established. Some editing will inevitably be necessary. This may simply involve changing the font, orientation or position of annotation for purely cosmetic reasons, but it may also involve a number of more complex operations. For example, intersecting lines may need tidying to ensure that they actually meet, corners of buildings may need to be squared.

If two or more map sheets form the basis of the same GIS, problems may occur at the edges of the maps even if they are based on a common referencing system. Line or area features which cross map boundaries will need to be correctly linked so that they appear as single features irrespective of map boundaries. Discrepancies may also arise through deficiencies in the mapping process, or through distortions in the mapping source. In this case a more complex process of stretching or compressing map edges using mathematical transformations is required; this is known as rubber sheeting, and is a feature of most GIS software.

Data may also need to be restructured. In order to ease the data capture process data may have been captured in simple spaghetti format and will need to be restructured into topological format.

Another restructuring requirement in a GIS is to convert between raster and vector formats. While vector to raster is relatively straightforward, there is inevitably some loss of information; the true mathematical line defined by vector co-ordinates may not coincide with the raster grid size and some displacement will occur. Raster to vector conversion can be semi-automatic with more software becoming available with line-following facilities. The software will attempt to follow line features in the raster data, creating vector data as it moves. Operator intervention is required if an intersection is located. Line snap features are also included to ensure that intersections are tidy, and no unwanted gaps are created. This feature is especially useful for contour maps. Most GIS software caters, to some extent, for raster/vector conversions.

3.7   Data quality

The term "data accuracy" is commonly used to refer to the closeness of an observation to its true position; the more accurate the data, the more expensive it is to collect. It is important not to quote a distance to a precision that is greater than its accuracy. For example, the length of the M1 motorway can be described in terms of miles. If the measurement were only accurate to the nearest mile, to describe it more precisely in yards feet and inches would imply an accuracy not present in the data. In a GIS it is generally not possible, or even necessary, to display or print the data with complete accuracy. There is therefore a need to define, at an early stage, the accuracy required of the data, i.e. how close to the true value the data must be. The ability of a GIS system to display truly accurate data is largely dictated by scale; at any given scale there is a critical distance within which two points can be regarded as identical.

All data needs to be maintained and geographic data are no different; the single most common defect of GIS data is the failure to ensure that data are kept current. The data capture exercise, even for relatively small systems, can be prolonged and this means that even before the data capture exercise is complete, some of the earliest captured data may be out of date. Buildings may have been built or demolished, roads may have been re-routed; in some cases features may appear to have moved simply because more accurate measurements have become available. The effect of all these changes on the effective use of the GIS will need to be considered carefully, and a proper maintenance strategy adopted.

Considerations

There are a number of other data quality characteristics that need to be considered to ensure that the data is of an acceptable standard. These include:

- Has all the necessary graphic and non-graphic data been captured?

- Is the data currently consistent throughout the data layers?

- Is the data sufficiently current for the requirement?

The quality of the data will decay with time, but may not affect the use to which the data is put. The maintenance strategy will define when the database NEEDS to be updated and what historic data are to be retained. This will affect the base data layer as well as user specific layers. All layers will be closely linked and a relatively small change in one layer may require equivalent changes in all other layers. This problem becomes easier if a topological data structure is in use; changes to co-ordinate information can be transparent to the application users. Nevertheless users will need to know which version of the data is in use. In order to maintain a clear distinction between updated data and non-updated data, some form of version management will be necessary.

3.8   Data management

A GIS system is likely to be shared by a number of individuals or departments within an organisation, all with a need to update and use specific sets of data. In common with all IT systems, a GIS needs an effective data management policy in order to protect the integrity of the database. In many instances the issues surrounding data management are identical with those for conventional IT systems: for example, appropriate documentation of data, applications and responsibilities. However GIS does present some unique problems.

Maintaining the integrity between data layers is crucial; for this reason specific responsibility for updating data layers needs to be specified. The base data layer should only be amended under strictly controlled conditions. Appropriate access controls will ensure that only owners can update the data, and only authorised users can access the data.

There are a number of legal issues surrounding the use of a GIS, which must be considered in a data management policy. If decisions are made, which are based on inaccurate information, and which affect a third party, an organisation may be legally liable for the consequences of that decision. The information that is held must therefore be adequate for its proposed use. A GIS may also bring together previously disparate information and by doing so may be able to identify an individual. If this is the case the database will need to be registered under the Data Protection Act, and can only be used for its defined purpose. There are circumstances whereby a post code may be considered personal information. The copyright requirements of data must also be considered; users of the GIS must not be able to infringe copyright requirements (see Chapter 5, Section 5.2).

There will also need to be rules for the plotting and displaying of information. For instance, an attempt to display data on a larger scale than is appropriate could be misleading.

3.9 References

[1] Antenucci J., Brown K., Croswell P., Kevany M., Archer H.

'Geographic Information Systems - A guide to the technology' 1991 Van Nostrand Reinhold, ISBN 0-0442-00756-6

[2] Cooper M.A.R.

'Fundamentals of Survey Measurement and Analysis' 1974
Crosby Lockwood and Staples, London.

[3] Dale P.F., McLaughlin J.D.

'Land Information Management'
Oxford University Press, ISBN 0-19-858405-9

[4] Maguire D., Goodchild M., Rhind D. (Editors)

'Geographic Information Systems' 1991,
Longman Scientific and Technical, ISBN 0-582-05661-6

[5] Maling D.H.

'The terminology of map projections' 1968
The International Yearbook of Cartography

[6] Maling D.H.

'Co-ordinate Systems and Map Projections' 1973
George Philip, London.

4 Purchasing considerations

4.1 Introduction

Much has been written on the specification and procurement of GIS hardware and software, and there are many case studies. This chapter reviews some of the literature and is intended for business, IT and project managers and purchasers.

Greatest benefits come from full integration of GIS as part of the IS strategy. This means the operational limits of the GIS must be defined at an early stage, and the necessary databases which will source the attribute data, whether paper-based or electronic, must be catalogued. Only by doing this will full use be made of the GIS.

4.2 System definition

Several guidelines have been suggested for how to develop and define the requirements of a GIS, and as yet these are still at a formative stage. SSADM, the government systems analysis and design methodology, has not been adapted to deal with GIS specifically, though it has been used for several geographical information implementations, such as the CSRWR (Computerised Street and Roadworks Register) at the Department of Transport. It is especially useful for the larger projects (true of SSADM in general), though the analysis techniques such as data flow diagramming and data structuring, themselves not specific to SSADM, are necessary in most system developments. The actual system design and implementation of GIS are usually undertaken by the vendor, because of the technical nature of the applications and the non-standard nature of much of the software.

A number of factors are, by consensus, key to the development of a successful GIS system. One is the necessity to prototype or pilot the system properly (even if the early system only serves a limited number of people) as a guide to implementation costs, information needs and user reaction. Prototyping tends to be a fairly quick implementation of the application to give an idea of, for example, the look and feel, and to give potential users a chance to experience the system. A pilot system is suitable for a larger project but is more costly, and involves implementing a part of the application and loading data with a view to assessing the feasibility of the system. In either case, it is important to realise that the system constructed at this stage will most likely have to be completely re-built. The use of application generators to adapt the user interface and menu options will facilitate the development of these early systems.

Other factors key to success are common to all computer projects, such as the need for management commitment, the need to find a champion to spearhead the project and the need to delineate the ownership of the data at an early stage.

The Local Government Management Board have defined a methodology [4] to be used for the definition and procurement of a GIS. Within what the LGMB manual calls a strategy, there are the following phases:

1. Map use study

2. User requirements study

3. Detailed needs evaluation

4. System specification

5. Invitation to tender

6. Evaluation of tenders

followed by implementation.

A L Clarke [3] has built a 4-stage GIS acquisition model incorporating the work of several other authors. This in particular, specifies a pilot study at the end of stage 1, the analysis of requirements. It also has two stages of costing, a cost benefit analysis and a cost effectiveness evaluation. J C Antenucci et al [1] also suggest a pilot study, but after the system purchase. However, all these methodologies agree on the basic progression from requirements analysis, fed from the user, to a requirements specification and implementation.

Many of the more common GIS products have application generators built in. The functionality of these tools can be of use, if suitable programmers can be found. Sometimes the application generators will be little more than report generators, but others allow the building of a user interface with, for example, menus, help screens and labelled functions.

A GIS procurement is subject to the usual rules and guidelines outlined in the CCTA Guide to Procurement within the Total Aquisition Process [2], including obligations under the relevant EC/GATT legislation concerning the advertisement and award of contracts above a certain financial threshold and the use of standards in procurement specifications.

The cost-benefit analysis of a GIS can be difficult. Hardware, software, data and staff costs should be calculable, but the benefits are often intangible, being mainly in terms of service improvement. It is important to establish what cost-recovery procedures, if any, are to be used. The Local Government Management Board has its own cost-benefit methodology, that has been implemented on a software package using a Lotus spreadsheet.

4.3   Supplier selection

It is estimated there are now some 1,000 suppliers providing products and services to the GIS market worldwide. These companies will have emerged via various routes into GIS; originally this arena was the preserve of smaller niche companies, many from the CAD market, but now most of the larger hardware suppliers have an interest.

There are several general considerations when choosing a supplier. While the basic requirement should be that the company can fulfil the system requirements, there are other aspects of the company which can be key to the success of a procurement. These are:

- Financial considerations
- Track record
- Technical considerations
- Strategic considerations

Most considerations are general for all procurements, some are particularly relevant to GIS.

4.3.1 Financial consideration

The company should have the financial strength and stability to undertake the contract with minimum risk. This implies the company itself is financially healthy - accounts should be examined for profitability and liquidity. If the company is not particularly stable, a parent or associated company can be used to underwrite the contract. Secondly, the contract should not be so large, compared to total company turnover, that the company is unduly reliant on a single contract.

The level of risk should be balanced against the type of contract. If the system is not of prime importance, or there are alternative vendors, or on-going support is not needed, the long term viability of a company may not be so important.

Financial assessment can be particularly important when dealing with GIS companies, as many of them are still quite small and the future of the GIS market is uncertain. Even if the company itself is profitable, the GIS operations will not be supported if they are not profitable.

> **Questions to ask the supplier**
>
> - Who owns the company, and what are its financial resources?
> - What is the sales turnover from GIS products and other sales revenue from GIS activities?
> - How are these revenues split between UK and world wide activities?

4.3.2 Track record

The track record of the company is a useful guide to the history and commitment of the company to GIS, as well as its ability to fulfil a requirement. Many of the companies will specialise in particular market sectors and target their applications accordingly. Two of the largest of these markets are local government and utilities.

Outline details of large contracts can usually be found in the press, but the company should be willing to provide some test sites for their products.

> **Questions to ask the supplier**
>
> • What reference sites can be quoted?
> • Are the systems at those sites similar to the applications currently being considered?

4.3.3 Technical considerations

There are examples of functional specifications available, and these are discussed in Section 4.5. However, the specification must do more than simply outline the system needed. As far as the company itself is concerned, it is important that it is technically competent to undertake the contract, whatever it may claim. Examining previous implementations is useful, but it is important to look closely at the technical backup being offered, especially for the large number of GIS software products that have not been developed in the UK and may therefore lack sufficiently competent technical support in this country. Several GIS companies have ISO9000 quality approval for their operations. While this is no guarantee of quality, it does suggest a commitment to quality.

> **Questions to ask the supplier**
>
> • What are the nature and extent of implementation services available?
> • What training services are provided?
> • What are the levels and nature of user support provided?
> • What are the levels and nature of system support provided?
> • What are the arrangements for hardware and software maintenance?

4.3.4 Strategic considerations

It is important to choose a GIS which will have as few proprietary components as possible. In this way one can avoid lock-in to a particular supplier (which tends to increase the cost of ownership), and provide more flexibility for the future. The most obvious open route is to choose a UNIX based GIS where a suitable one is available, but even this may not ensure hardware independence.

The company's commitment to GIS and its future intentions should be considered. The company may have a policy statement on GIS and the direction it intends to take in product development. Some companies are actively involved in standards issues via the AGI and membership of the AGI suggests commitment to GIS.

Standards are slowly evolving in the GIS arena. The Digital Geographic Information Working Group (DGIWG), which comprises military representation from NATO countries, has agreed a data exchange standard, DIGEST. This has been adopted as the Military Survey exchange standard. GIS standards in the UK are co-ordinated by BSI/IST/36, and in Europe by CEN TC287. There is no ISO GIS standards committee. The only enshrined standards are for the interchange of data, most notably, in the UK, the NTF (National Transfer Format), BS7567. However, as with many standards, the definition is not exhaustive, so though data may conform to NTF, it cannot necessarily be read by an NTF conformant package. Work has been done on the SQL3 standard which will ensure better support for GIS, but this will not be issued until 1995. Further work needs to be done on GUI, benchmarking, system management and a common API, and this work is being undertaken by the European standards body CEN and by BSI (though the BSI standards are being developed by the AGI).

Questions to ask the supplier

- What are the supplier's medium and long-term technical plans?
- What standard does the product currently support, and which are planned?
- Which system components are not within the supplier's direct control?

4.4   Consultancy selection   As stated in Section 4.1, installing a GIS will have implications for the IS organisation. The pros and cons of installing a GIS will need to be made clear to senior management, to gain their commitment, while independent consultants with a wide experience of IS strategies and GIS may be needed for a strategy study.

At various stages in the project lifecycle, it may be necessary to utilise different sets of consultants and/or service suppliers. Consultants may be members of the RICS (Royal Institute of Chartered Surveyors) or the BCS (British Computer Society). The Land and Hydrographic Survey Division (LHSD) of the RICS gives professional certification for GIS practitioners. However, not all division members will be GIS specialists. The LHSD will be able to identify GIS specialists. The BCS gives professional entry for general computer specialists and has a GIS interest group. However, membership of this group does not imply professional membership of BCS.

In the early stages of a project, there may be a need to hire project managers and systems analysts, for a fixed term, to run or assist in the project (see the CCTA IS Guide E2 [2] on the hire and management of consultants).

At the data input stage, there may be an additional need for a bureau service. There are several agencies that can provide personnel for these tasks, which tend to need quite specific skills related to the digitising of data, such as hand-to-eye co-ordination, attention to detail and ability to do repetitive tasks. Manpower, who provide staff for data input, have laid down criteria for GIS staff selection along these lines.

As with the suppliers, it is important to realise that the size of the consultancy or service provider is not a good measure of expertise, indeed some of the smaller consultancies will have far greater experience in terms of numbers of consultants actively working in GIS, and the number of years experience.

> **Questions to ask the consultant**
>
> What is the consultant's experience in the following areas, where applicable?
>
> - GIS strategy
> - Geographic information needs assessment
> - Cost/benefit analysis
> - Technical & financial feasibility
> - User/Operational Requirements specification
> - Systems selection
> - Procurement management
> - Data conversion management
> - Implementation management
> - Project management
> - Configuration/capacity planning
> - Business analysis
> - Technical/project audit
> - Training
> - Market research
> - PRINCE
> - SSADM
> - Total Acquisition Process (TAP) [6]

4.5   Factors in system selection

Refer to the Local Government Management Board's (LGMB) GIS Functional Specification [5] for a set of system selection criteria. This is an exhaustive document. The LGMB is currently preparing a base specification and a set of application-specific specifications. Consultancy advice can be sought to check the invitation to tender and functional specification documents.

A functional specification may divide the system components into:

- Data collection/input

- Map management

- Storage

- Data manipulation and spatial analysis

- Output

- Performance and security.

Many software packages nowadays are built in a modular form, which allows the purchase of various options depending on need. This makes the purchasing considerably easier and means that one is not encumbered with an excess of functionality which will not be used. Similarly, some packages are available for machines of various sizes, so that cut-down versions will be available for smaller systems while keeping inter-operability with larger platforms.

The following sections outline points to consider. The list is not exhaustive.

4.5.1   Data collection/input

Chapter 5, Section 5.1 discusses the various sources of geographic data that are available. The application may need to import data from one or more of these sources. The most common form of background or foreground data is Ordnance Survey (OS) maps. Note that there is a separate Ordnance Survey for Great Britain and for Northern Ireland. Most data from the OS and OSNI is in NTF v1.1 format. From April 1993, OS will issue its data to its preferred format (BS7567 Part 2), and data in the older, DXF or OSTF formats, will be phased out. Note also that Military Survey is responsible for the provision of geographic data to MOD users.

In the past, OS data has been available in structured and unstructured form. Unstructured data consists of lines and is 'uncleaned', meaning the lines do not necessarily meet. Structured data, the form of geographical data now sold by the OS, has been cleaned (the lines meet at nodes) and buildings have seed points ie a pair of co-ordinates used to represent the area. Soon it is hoped object-based data will be available from the OS. It should be noted that OS data is not faultless, so an ability to edit it is useful. An ability to read the OS feature codes defining various physical structures is necessary. If a number of OS maps are being used, the ability to do edgematching or 'rubber sheeting', the fitting together of OS maps, may be useful, as the map edges do not always coincide. There are various ways of doing rubber sheeting, either automatically, using point or spline fitting, or manually. It may be useful, especially in terms of storage saving, to be able to fit curves as well. All OSNI data is fully structured and edge-matched.

A common form of data is raster images, which can generally be input from scanners. Raster data can be used as background data to aid visual interpretation of user data. Alternatively, raster data can have an intrinsic value. For example, satellite data might represent change in vegetation cover. Data can be scanned automatically and fairly cheaply and the price of scanners has been falling. The application may allow the transformation of vector data to raster form (note that raster data is also covered by the BS7567 transfer standard). Raster data may have editing tools - though this is not common. Perhaps more common is the on-screen digitising of raster data (Chapter 3 Section 3.6). It may also be useful to be able to attach attribute data to specified points on raster maps.

Making a gazetteer requires the creation of a list of names eg. streets, combined with a unique identifier and spatial location information. The gazetteer is used to allow access to data via this list. Most of the more common packages can use data from a range of the more common commercial database management systems such as Oracle, Ingres or Rdb. Some GIS use one database for both the textual and graphical data, while others have separate databases for these data sets, either of which may be proprietary.

Packages should have panning and zooming facilities which will allow the user to navigate around maps. If vector and raster maps are displayed simultaneously, they will undergo visual transformations.

Questions to ask the supplier

Concerning data formats and conversion:

- Which data exchange formats are supported?
- What data conversion facilities and services are provided?
- How much data translation/cleaning/re-structuring is expected?
- How does the system access information from databases on other systems?
- With which DBMSs can the system communicate?
- Whether read and/or write access is possible?
- How will the system share informis ation with remote sites?

4.5.2   Map management

Once the map data has been input, there must be a method for updating it; out of date data is a potential hazard. As yet, change-only update of OS maps is not available, and the entire map has to be re-input. Updating is a large and difficult task, prone to human error and which should therefore be automated as much as possible. OSNI data has provision for change-only update.

Recorded with maps must be such details as reference number, scale and version number. The maps may need to be accessed in a variety of ways, perhaps by street, town, or postcode.

Vector data can be stored in a number of ways, most simply as a collection of co-ordinates, but it may be better to have a topological database, where shapes are stored with their relationships. However, polygonal databases tend to be more expensive than a straightforward 'spaghetti' storage, if only because more storage is required.

Layering, where different layers can be treated independently, is particularly important for spaghetti data, or for topological data if there is a large amount of unrelated, or not logically connected, attribute data. Alternatively, layering can be used to compare maps or data.

For the user map data, many of the same concerns arise as with OS data ie. the checking of consistency or choice of input and output formats. Geographical data can be input in two forms: raster or vector, the former by scanner, the latter by digitising tables and field data capture, including satellite positioning systems [1].

Questions to ask the supplier

What are the facilities for:

- Managing and editing base maps?
- Managing and editing fore-ground features?
- Image handling?

4.5.3 Storage

The improvement in the performance and price of storage systems has improved in the last few years, which is advantageous as geographic information uses up large amounts of storage. There are various techniques of data compression that will allow the storage of geo-graphical information more economically (see Chapter 3 Section 3.3).

Questions to ask the supplier

- What is the role of data storage and transfer devices?
- How were storage capacities calculated?

4.5.4 Manipulation and analysis

Some of the more common and useful functions available in GIS are: line, area and perimeter calculation; merging, overlaying and intersecting polygons; weighting and summing spatially co-incident values; spatial and trend analysis; and tracing and computing network flows. On the attribute data itself, the package may have facilities for data retrieval, using spatial delimeters (such as 'within a specified window' or 'within a specified distance from a point or line'), and Boolean operators (AND, OR, XOR, NOT). The package may be able to display or tabulate this data where particular conditions are satisfied. In some systems, aggregate objects (such as houses or roads) can be defined and manipulated.

For the more commonly used manipulations and analyses, the user should be able to set up menu systems where the calculation will be done automatically.

Questions to ask the supplier

What facilities are provided for:

- Spatial analysis?
- Application building?

4.5.5 Output

As with storage, the cost of a given performance level of graphics terminals has reduced markedly in recent years, and should continue to do so. Ergonomics will determine the screen size and resolution; PCs have in the past tended to be either EGA or VGA, but are increasingly providing resolution comparable with workstations. Similarly, a range of screen sizes are available.

The user interface may be proprietary or standard (eg. Windows or presentation manager for PCs, Motif or OpenLook for workstations). The package may allow the user access to the full range of GUI functions, such as multiple windowing, window sizing, multi-tasking (more a function of the operating system).

For the printers, plotters and terminal screen, it is necessary to have enough colours, line styles, text styles, for the appearance to be clear. In terms of the relationship between printer and screen, it may be necessary to have a screen print, or a user definable print, or both. Also, the print may need title bars, map scale, version number, grids, and directional pointers.

Map plotters are very expensive (electrostatic and ink-jet being the most common). They have the advantage of speed over printers. However, printers have the advantage of being able to print both vector and raster data, while most plotters can only cope with the former. The process of plotting can be considerably quickened if the plotter has the ability to process intelligently the plot. This is because a plot may not be digitised in a logical sequence, and the number of pen movements can be considerably reduced by sequencing the pen movements.

Report writers are essential with most systems, though this may be a separate module in some GIS. Other features which could be useful are the ability to pre-programme some mathematical calculation, preferably in SQL, and the ability to create menus.

Questions to ask the supplier

What are the facilities for:

- Screen, menu and symbol design?
- Plot layout design?
- Report layout design?

What interfaces are provided for printing and plotting?

4.5.6 Performance and security

All parts of the data handling process - data input, screen drawing, querying, plotting, will have their own performance requirements. The speed may be defined by the hardware or the software, though some processes naturally take longer - for example, displaying (raster) data on a screen is a slow operation without a fast graphics card or co-processor. In a distributed system, there may be a need for a high speed network to transfer large amounts of data.

Levels of security can be provided by the database or the application itself. Security requirement may be complex for a GIS, as there may be a need for spatial security, for example around sensitive sites, as well as various levels of attribute security.

Questions to ask the supplier

- Ask the supplier to demonstrate the performance of the system, particularly where this may require remote access to geographic data?
- On what basis were the processing/communications units proposed?
- Will the system provide adequate performance?
- How will the system perform in the event of a power or system failure?
- How can the integrity of data be restored after such a failure?
- What features prevent unauthorised access to the system or specific datasets?

4.6   References

[1] Antenucci J., Brown K., Croswell P., Kevany M., Archer H.

'Geographic Information Systems - A guide to the technology' 1991
Van Nostrand Reinhold
ISBN 0-442-00756-6

[2] CCTA

'The Information Systems Guides: E2 The hire and management of consultants' 1989.
John Wiley and Sons Ltd
ISBN 0-471-92545-4

[3] Clarke A.

'GIS specification, evaluation and implementation'

Geographic Information Systems' 1991 Volume 1 Pg 477-88.
Longman Scientific and Technical ISBN 0-582-05661-6

[4] Local Government Management Board

'An approach to evaluating Geographical Information Systems for Local Authorities' 1989
ISBN 0-7488-0048-4

[5] Local Government Management Board

'Local Government Geographical Information Systems Product Software Functional Requirements Specification' 1991
ISBN 0-7488-9876X

LGMB
41 Belgrave Square
LONDON
SW1X 8NZ
071-235-6081

[6] CCTA

A Guide to Procurement within the Total Acquisition Process 1991.

5 Information purchasing considerations

5.1 Sources

5.1.1 Data requirements

Before looking for the possible sources of geographic data it is essential to identify the exact type of data required. For example, if the system is to record the structure of a road network it may not be necessary to have details of items between the roads; eg. buildings, railways, etc. Failure to make this identification can result in higher purchase costs, higher data maintenance costs, larger sized data files and performance problems.

In looking at land surface information, mapping organisations will provide digital map data which includes the land features at a particular date. These data may not be the last time a survey was done, as the mapping organisations may only release new versions of maps when a significant number of features have changed. For example, it is feasible for a map not to show something as major as a new motorway junction as it could only count as a single feature change.

Deciding on the features required is also important; current digital map data only has a subset of the features used on paper maps. If it is essential to have some of the omitted features, they will have to be added by the purchasing organisation.

It is also possible that for a particular system the majority of the data is unnecessary, and/or it is critical to have the data accurate to a particular level of currency (see Chapter 3 Section 3.7).

There are many suppliers of geographic data; however, for the reasons identified above it is also possible that the most cost effective solution for an organisation is to gather the raw data itself.

A large number of companies will produce digital data from aerial photographs, satellite images and ground surveys. In addition to having the appropriate data for the system, there are a number of other possible advantages, including:

- the updating and maintenance of the data, including frequency and cost, is wholly in the organisation's control

- the cost of collecting the data could be offset, or even fully recovered, by selling it to other organisations (as long as the ownership of the data has been stated in the original contract - see Section 5.2.2).

> **Questions to ask the supplier**
>
> Concerning the data provided:
>
> - What is the content?
> - What are the features?
> - How complete is it?
> - What is its currency?
> - How frequently is it isssued?
> - How it data integrity maintained?
> - How is the quality of the data controlled and managed?

5.1.2 Types of data

Geographic information is not necessarily of a map type format; it can contain graphic and/or non-graphic data.

Graphic data can be in many forms, including for example:

- Digitised photographs

- Satellite images

- Full land feature maps

- Street maps

- Underground services maps

- Hydrographic charts

- Diagrammatic maps (eg. London Underground).

Non-graphic data comes in an even larger variety. It can be in virtually any form of relational or tabular database; and consist of any information that is an attribute to a specific location (see Chapter 3 Section 3.1). For example, the address details of a particular building, its maintenance records and the personnel records of the staff working within it could be in three different databases, all of which could be used as the non-graphic data in a geographical information system about the organisation's locations.

Much of this type of non-graphic data is already held by an organisation, but there are other types (for example census and other socio-economic data) that are available on the open market.

> **Questions to ask the supplier**
>
> Concerning the types of data supplied:
>
> - What coordinate/map projection system is used?
> - How is the data structured?
> - In what format(s) is the data delivered?
> - How many records are involved?
> - How large are the data files?
> - On what media would the data be delivered?

5.1.3   Types of source

There are suppliers at all stages in the development of geographic information: from survey companies which can provide raw data; through suppliers which have organised the raw data, perhaps into maps; value added data suppliers which have combined and enhanced a selection of data sets; to geographic information systems user organisations which sell their data.

Whether or not an organisation is buying in the geographic information or developing its own, it will almost certainly be utilising a number of sources in order to have all the information it requires.

As a result, there are an increasing number of companies which provide value added data services. These companies take raw data from a number of sources, combining it, possibly amending it and often adding their own data, and then reselling it. In the UK this market is still relatively small, and its long term future will depend on the level of growth of the overall GIS market and the availability of sufficient data sets.

Defence users of geographic data should use Military Survey for its provision.

Sources of geographic data are found in both the public and private sectors. Map data, for example, is available from specialist mapping organisations such as Ordnance Survey, Ordnance Survey of Northern Ireland and Bartholomew, but also from organisations like the Automobile Association which sell the data developed for its own geographic information systems.

Organisations such as Local Authorities and the utilities are currently investigating the possibilities of marketing their data.

In central government, there is the Tradeable Information Initiative (TII) which is run by the Department of Trade and Industry (DTI). The TII is intended to bring to the market that information which has a commercial value and is held by government departments, in order to offset the costs of its production.

Two organisations have extensive experience in helping government departments achieve the aims of the TII, namely Taywood Data Graphics and The Electronic Publishing Services. Taywood Data Graphics' experience of tradeable information has included geographic information with the Ordnance Survey and English Nature

---

Questions to ask the supplier

Concerning data sources:

- What was the source?
- What was the scale from which the data was derived?
- What was the nature of the conversion process?
- What is the spatial accuracy of the data?

5.2 Costs of purchase

5.2.1 General issues

As with most electronic data and applications, the purchase of geographic data is controlled by a licence issued by the owner. Ownership of data and applications rarely passes to the purchaser, instead the owner issues the licence to allow the purchaser to use the data or application within a set of constraints. This is based on the concept of intellectual property rights as contained in copyright legislation, which implies that copying is equivalent to theft where it deprives the owner of income.

The legal implications of a GIS must be considered if information on living individuals is to be kept. The Data Protection Act (1984) specifies the conditions under which such data may be kept, and what it may be used for.

Where an organisation makes decisions based on data it has bought but which subsequently turns out to be inaccurate, the question of legal liability may arise particularly if the decision affected a third party. Consequently, the initial purchase contract should clearly state where legal liability for the accuracy of the data rests.

5.2.2 Legal position

The copyright laws vary between countries, although the EC is currently developing legislation for harmonisation. In the UK, copyright is automatic on any document or equivalent provided that it is original (ie, not a copy), whereas in many other countries copyright only applies where there has been some intellectual input in the production. For this reason, in some countries, maps may not be protected by copyright because they can be regarded simply as a record of a physical situation.

The UK's Copyright, Designs and Patents Act of 1988 includes specific provision for the different requirements of electronic works compared with paper works. The Act defines the types of property covered by copyright, the actions that are equivalent to copying, and makes copying without the owner's permission illegal. The copyright of applications software and the conversion of paper to digital formats are protected very specifically in the 1988 Act. However, digital data is only covered in the more general class of literary works.

The issue of copyright is avoided where an organisation is the owner of the data, for example where it has commissioned the initial survey work; and, of course, ensured the issue of ownership is detailed in the contract with the surveyor.

Questions to ask the supplier

Do the terms by which the data is available include the right to:

- Make backups?
- Make other copies for internal use?
- Produce prints or plots?
- To own derived data?

What is the liability of the supplier company?

Does the supply of the data constitute an invasion of privacy?

5.2.3 Suppliers' position

The suppliers calculate their charges and develop their licence agreements based on the Copyright, Designs and Patents Act 1988, and generally offer one of two options.

With simple, often PC based, types of geographic data, the market is moving towards the 'shrink wrapped' type product. The purchaser pays a single fee, which provides them with a copy of the data and a lifetime licence to use it.

With other more complex data the suppliers have a charging structure made up of either three or four components: purchase, maintenance, copying and in a small number of cases, usage.

Purchase

As with the 'shrink wrapped' pricing, this is a single fee which provides a copy (or copies) of the data under a licence to use.

Maintenance

An agreed fee to cover the cost of developing and providing updates in the future, usually based on a service level agreement. The fee will be based on the frequency of update and the anticipated number of changes to the product.

The updating of maps is usually triggered in one of two ways; units of change or regular time intervals.

As features alter, changes are made to the mapping organisation's base map. When the agreed number of feature changes has happened (often in the range of 15 to 20 features) the mapping organisation's working map is updated from the base map and a version of the new working map is issued to the customer. There are two potential main problems with this method: either there are too few changes to trigger a new version even though one of the changes is to a feature critical to the organisation; or, there are many small changes that are insignificant to the organisation but which keep triggering new versions of the map.

The alternative is to receive a new version of the map at a regular, and specified, time interval. This ensures that the organisation receives all feature changes within a maximum, known time period. But the new versions may only contain either a small number of changes, or changes that are insignificant to the organisation.

The maintenance fee will of course be calculated by the mapping organisation on the amount of resource it will use in meeting the service level agreement.

Copying fee

A fee will normally be levelled for each hard (paper) copy taken of the map data. For example, a utility company might print a map each day for each of its engineers marked with the details of the sites and work to be carried out that day. This allows the copying fee to be calculated on expected demand and then possibly checked by audit.

Usage

Duplicating an application held on one hard disk on to another disk is clearly copying. But is duplicating an application held on a hard disk into working, or transient, memory also copying?

For software applications, the Copyright, Designs and Patents Act of 1988 makes clear that copying which is incidental to use is not a breach of copyright. However, data does not have a similar clause for copying incidental to use in the 1988 Act. Some organisations, most notably the Ordnance Survey, claim that this allows them to charge a usage fee for each time a piece of their data is loaded into working memory from storage.

Whilst this may be a correct interpretation of the Act, it is without precedent, and would need a test case to clarify the position. In the meantime, purchasers of geographic information from the Ordnance Survey should be aware that there will be a charge made for 'usage' based on the Ordnance Survey's estimate of the use of its data (eg. the number of times the data is loaded into working memory).

While copyright control on Ordnance Survey of Northern Ireland's data is generally similar, implementation does not not necessarily follow identical rules to Ordnance Survey.

Questions to ask the supplier

What are the charges for:

- Purchase of data files?
- Maintenance?
- Plotting and printing?
- Usage?

How stable are the charges?

What is the pricing structure and what discounts are there?

5.3   References

[1] Larner A.

'Digital Maps - What you see is not what you get'
Land and Hydrographic Survey - March 1992

[2] Larner A.

'Encouraging Data Market Growth'
Mapping Awareness - November 1992.

6 The GIS market

6.1 Market sector shares

Introduction

There is a variety of GIS software and hardware incorporating technologies for data capture, management, processing and analysis. GIS products provide some or all of these functions. GIS covers a number of markets and its current state of development is difficult to quantify.

There has not been a comprehensive 'definitive' study undertaken to establish the size of the GIS market and to identify related segmental details which would allow a truly credible comparative analysis. Estimates are therefore somewhat speculative and can vary considerably depending on the definition of GIS used and the year upon which projections are based. Most surveys show market share by revenue; this has become more misleading with the advent of lower priced hardware and software. A more accurate measure would be the penetration of installed GIS and the number of pilot systems. However, such data are scarce and more difficult to compile.

Companies worldwide are beginning to exploit the benefits of GIS. Although it is difficult to identify the penetration of GIS in business applications and the use of general purpose systems for GIS related work, there are signs that business take-up is increasing. The take-up is more established in government (in the UK, mainly local authorities) and public utilities where the applications are more visible.

This chapter largely draws information from Dataquest, a market research company. Dataquest includes GIS as a segment of the CAD/CAM/CAE market. The other segments are: mechanical systems, architecture, engineering and construction systems; and electronic design automation systems. Dataquest defines a GIS as a computer based technology composed of hardware, software, and data, used to capture, edit, display and analyse spatial information tagged by location.

ms

Figures 6.1a and 6.1b show the revenue breakdown of
the world CAD/CAM/CAE market in 1991. GIS has
become the fastest growing segment within this field;
existing CAD facilities can be used as a vehicle for
developing simple GIS applications as many of the
requirements are similar.

Despite poor economic conditions, the total world CAD/
CAM/CAE market grew by 9.5% to $15.8 billion in 1991
(Dataquest [4]). The market is forecast to grow at a
compound annual rate of 9% through to 1996.``

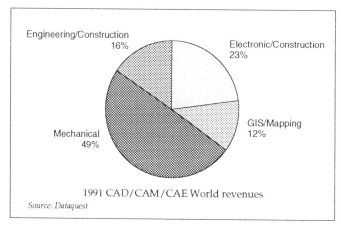

Figure 6.1a: GIS market shares.

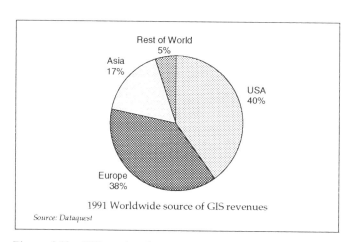

Figure 6.1b: GIS market shares.

Global market

Figures 6.1a and 6.1b also show the estimated worldwide distribution of total GIS revenues in 1991. Taking a broad definition of GIS to include turnkey systems and related services, a consentient view suggests that the global market generates annual revenues of approximately $2 billion. Dataquest estimated revenues were $1.98 billion in 1991 and would be $2.33 billion in 1992. In 1991, hardware accounted for 53%, software 30% and services 17%.

USA

North America has been dominant in the development and implementation of GIS although recent data suggests that the US segment of the world GIS market is only marginally ahead of Europe's share. At the end of 1991, the US General Accounting Office (GAO) reported that the number of government agencies using GIS had increased from 18 to 44 since 1990. However, significant cuts in the defence budget over the next five years have been proposed by the US Administration. Companies that are heavily dependent on government defence contracts will have increasingly limited growth opportunities.

Europe

The take-up in Europe is approximately $800 million. According to Green [6] in 1990, the three market leaders in Europe were the UK, Germany and Italy (based on 1989 statistics). Figure 6.2 shows the European GIS market shares in 1991.

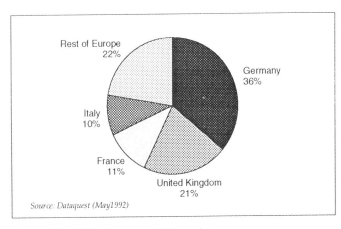

Figure 6.2: 1991 European GIS market.

UK

The GIS market in the UK is immature, but arguably more advanced than in any other European country, at least in terms of digital map availability. Projects tend to be smaller in geographic scope but more detailed than in the USA. The annual revenue derived by suppliers is approximately £120 million.

A recent survey by the University of Sheffield Department of Town and Regional Planning [2] shows that only 16.5% of the UK's 514 local authorities have acquired a GIS and most of these are still in the pilot phase. This is surprising, given that approximately 80% of local authority information refers to a specific geographic location. The reasons given for the relatively low take-up were:

- a lack of management commitment to GIS within some authorities;

- a lack of financial resources;

- some authorities felt that GIS facilities were too expensive;

- a lack of GIS experience within authorities.

The commercial take-up in the UK is becoming more conspicuous in a number of industry sectors. Private investment has also been boosted by the recent privatisation of many utilities.

Australia

The 'rest of the world' category in Figure 6.1b includes Australia where there has been a considerable take-up of GIS in large utilities and state governments [1]. The Australia New Zealand Land Information Council (ANZLIC) is supporting the development of the Australian Spatial Data Transfer Standard (which will be similar to the standard in the United States) and more effective management and use of land information. A major project, undertaken by the National Resource Information Centre, has been the development of a regional scale GIS for the Murray-Darling Basin Commission.

6.2   Future market trends

GIS is a young technology and the industry outside the academic sector has not expanded as quickly as market analysts had expected. Estimates vary but the general view is that the annual growth rate is currently about 20% (10% for CAD/CAM/CAE according to Dataquest). Although such projections can be misleading given the varying combinations of GIS technology, the market is considered to be a key growth area and is attracting an increasing number of vendors of both GIS software and GIS hardware platforms. On the demand side, GIS will be used more to identify new business opportunities.

Dataquest forecasts that the GIS market covering all platforms will be worth approximately $4 billion in 1995. GIS revenues derived in Europe are forecast by Dataquest to grow more rapidly than in the US. Growth forecasts generally have recently been scaled down because of the recession and the impact of reduced expenditure plans by governments.

Positive influences

The following factors will induce market expansion:

- hardware performance and data storage capacity have increased - most GIS were confined to mainframe systems; an increasing number run on workstations and PCs

- workstation costs are falling

- increasing availability of application specific software at lower prices

- stabilisation of standards for GIS

- many organisations have data that could be utilised by a GIS

- sophisticated relational database software is making data manipulation easier

- the availability of digital data (although this is relatively expensive)

- the supply of GIS services is increasing

- an increasing number of government agencies and corporations are recognising the significant potential benefits of GIS (eg. land resource planning and market analysis)

- the potential for co-ordinated GIS sharing

- increasing need for statutory registers (particularly environmental such as contaminated land registers).

Negative influences

Factors restricting growth are:

- current pressures on governments to reduce public expenditure

- many GIS applications are still stand-alone limited user systems operating in specific parts of organisations

- systems still tend to be proprietary and expensive and implementation can be costly and time-consuming

- the high cost of data capture, conversion, updating and management (generally estimated to be 80% of total GIS costs)

- problems of integrating data from a number of disparate systems (GIS and non-GIS)

- there is both a demand and supply side management resistance to risking GIS investment - difficulty in quantifying cost/benefits (benefits tend to be long term)

- pilot tests and prototypes may not live up to expectations - often due to costs (although pilots can identify hidden benefits)

- a successful GIS implementation requires good project management; this often is not considered.

- unsuccessful implementations restrict market expansion

- legal concerns (particulary copyright protection of data).

The prospects

There is a widespread applicability of GIS, and despite factors that can restrict its growth, the market will rapidly develop as real costs reduce further and the level of acceptance and use gains momentum. Technology, per se, is not an obstacle to the expansion of this market although the integration of GIS with conventional information systems, expert systems and multimedia through the development of IT standards (such as interchange formats), would further enhance applications potential. Business take-up is likely to increase as the systems become more cost effective and easier to use, and managers become more aware of the long term benefits (therefore implementing the appropriate IS strategy). A GIS does not necessarily imply information pertaining to land or maps in the conventional sense. It is perfectly feasibile to have a GIS relating to a building or an aircraft, for example. GIS will become a corporate resource rather than a departmental pilot. Longer term, it is likely that there will be a 'shakeout' of suppliers as the market matures further. The term GIS may disappear as the technology becomes assimilated into mainstream computing.

6.3 References

[1] Australian Land Information Council

'Land Information Management in Australia - Status Report 1988-1990' (1990)
'National Strategy on Land Information Management' (1990)
'1990-91 Annual Report' (1991)

[2] Campbell H. and Masser I.

'The Impact of GIS on Local Government in Great Britain' (1991)
Department of Town and Regional Planning (University of Sheffield)

[3] Dangermond J.

'The Commercial Setting of GIS'
'Geographical Information Systems' 1991 Volume 1 P55-65
Longman Scientific and Technical ISBN 0-582-05661-6

[4] Dataquest Market Statistics

CAD/CAM/CAE & GIS applications (May 1992)
Dataquest Inc, San Jose CA 95131-2398

[5] Gantz J.

'Coming of Age'
Computer Graphics World - October 1991

[6] Green R.

'Geographical Information Systems in Europe'
The Cartographic Journal, Volume 27 June 1990

[7] IBM

'Directions in Geographic Information' Volume 2 Issue 3 (1991)

[8] Maguire D., Goodchild M., Rhind D. (Editors)

'Geographical Information Systems' 1991
Longman Scientific and Technical ISBN 0-582-05661-6

[9] 'Mapping Awareness and GIS in Europe'

Published by Miles Arnold ISSN 0954-7126

[10] UNICOM

Greener A., Hart A., Pearson E., Tulip A.
'Geographic Information Systems'
Draft Information Technology Report 1992.

Glossary

The following definitions are to aid the reader's understanding of terms used in this Guide. A more comprehensive dictionary of GIS terms is produced by the Association for Geographic Information (AGI). The AGI dictionary is intended to provided a basis for terms used within the procurement process.

2.5D

A system in which the third dimension is constrained to a very simple relationship with the other two dimensions (eg where Z is a single valued function of X and Y).

Acid Test Ratio

The ratio shows liquid assets (debtors and cash) against all current liabilities (everything due within 12 months of the balance sheet date).

Address matching

Relating street addresses to point locations or areas such as census blocks or the location of buildings or emergency response incidents.

AGI

The Association for Geographic Information. The UK umbrella organisation for geographic information and its associated technology.

Annotation

The alphanumeric text or labels on a map, such as street or place names.

ANSI

American National Standards Institute - the US national standards body.

API

Application Programming Interface - the formally defined way in which an application program can interact with an operating system or other system resource.

Arc

A locus of points that forms a curve which is defined by a mathematical function.

Area

A bounded continuous two dimensional object which may include its boundary. Usually defined in terms of an external polygon or of a set of grid cells.

Attribute	A type of non-graphic data which describes spatially referenced entities represented by graphic elements.
Benchmark	A defined workload against which a computer's performance may be measured.
Bit	A logical two-way switch, capable of taking either of the values 0 and 1.
BS 7567	The British Standard for the transfer of digital geographic information (see NTF).
Buffer generation	Ability to calculate and generate an area around an object.
CAD	Computer Aided Design.
Cadastral survey	A survey based on the measurement and marking of boundaries for the registration of land ownership.
CAE	Computer Aided Engineering.
CAM	Computer Aided Mapping.
Cartography	The organisation and communication of geographically related information in either digital or non-digital form.
CISC	Complex Instruction Set Computer. The conventional instruction set which evolved to satisfy the needs of system software to enable the generation of reliable and efficient object code, often in a time-sharing environment governed by the operating system.
Client server model	Describes an IT configuration where an application uses the processing power of both a personal computer and a host system. The personal computer provides an interactive user interface and the host provides large-scale data storage and multi-user information sharing facilities.
CON29	Local authority control form for capturing land and property information.
Currency	The level to which data are kept up to date.

Data capture	The creation of digital data from existing information sources. In the context of digital mapping, this includes digitising, direct recording by electronic survey instruments and the encoding of text attributes.
Data transfer	Transfer of data between GIS systems (requires a transfer format independent of any GIS' internal data structure).
Database	A structured organisation of records, for purposes such as automatically generating up to date reports, and answering ad-hoc queries.
DBMS	A software system designed for creating, updating and retrieving information from a computer database; the software automatically manages the storage and processing of the data comprising the database.
Device Driver	A program providing an interface between the operating system and application software to support input from or output to a specific peripheral device. It translates commands issued through application software into instructions that a peripheral device, such as a plotter, can interpret to perform a certain function.
Digital map data	The digital data required for the user to create a map on the screen or create a hard copy map.
Digitising	The conversion of analogue maps and other sources to a computer readable form. This may be point digitising, where points are only recorded when pointing the cursor and pushing appropriate buttons, or stream digitising where points are recorded automatically at pre-set intervals of either distance or time. See also scanning and vector.
DTM	Digital Terrain Model. A digital representation of relief (ground surface). Usually a set of elevation values in correspondence with a grid cell.
DXF	An AutoCAD system dependent geographical data transfer format.

Edge match	The process of ensuring that data along the adjacent edges of map sheets or some other unit of storage, matches in both positional and attribute terms.
Feature code	An alphanumeric code which describes and/or classifies geographic features.
Functional requirement	Specification of a business task or process that a system or system component must perform.
Gearing	Refers to the extent to which costs have been financed by borrowing. A company is said to be highly geared when it has a high ratio of borrowing to shareholders funds.
Geocode	A code which represents the spatial characteristics of an entity. For example, a postcode or pair of grid co-ordinates.
GIS	Geographic Information System. A system for capturing, storing, checking, integrating, manipulating, analysing and displaying data which are spatially referenced to the Earth.
GPS	Global Positioning System. A constellation of US satellites. The satellites transmit signals which can be decoded by receivers to determine positions anywhere in the world.
Graphic data	Data which can be used to create a graphical representation of objects in the real world, eg. maps, plans, charts, photographs.
GUI	Graphical User Interface. A user interface which makes use of graphical objects, such as icons, for selection of options, and usually has a windowing capability, enabling multiple window displays on the same screen.
Layer	A subset of digital map data selected on a basis other than position. For example, one layer might consist of all features relating to roads and another to buildings.

Line	A one dimensional object. A line segment is a direct line between two points. A line is a series of consecutive line segments with common attributes.
Liquidity	The ability of a company to pay amounts owing as they fall due for payment; a measure of the degree to which current assets can be used to pay off current debt.
LLC1	Local authority form for local land charges.
Motif	Motif is a graphical user interface developed by the OSF for UNIX systems.
Multimedia	The use of text, data, still and motion video, sound, and computer graphics by a program to form a composite display.
Network tracing	The ability to locate a route through a network.
Node	The start or end of a link or line. A node can be shared by several lines.
Non-graphic data	Textual data or attributes. Digital representations of the characteristics, qualities, or relationships of map features or geographic locations.
NTF	National Transfer Format. A UK standard for the transfer of geographic data, administered by the AGI. It subsequently evolved into a British standard (see BS7567).
Object	A collection of entities which form a higher level entity within a specific data model.
Object-oriented programming	The writing of computer programs using object-oriented techniques and languages. These employ a data-centred approach to programming, based on the definition of 'objects'.

Operating system

The innate controlling software of a computer. Among its functions is the management of resources: for example, allocating memory and processor time to applications and other processes running in the machine. The operating system is also responsible for transferring data between the storage media and the memory.

Optical disc

A compact data storage device consisting of a disc whose coating can be altered to encode information. Data can be read from the disc by laser light which is reflected from the surface according to the new property of the coating.

OS

Ordnance Survey (Great Britain).

OSF

The Open Software Foundation was formed in 1988 by IBM, DEC, Hewlett-Packard, Apollo, Siemens, Nixdorf, and Groupe Bull to develop a version of UNIX as an alternative to AT&T's UNIX System V (SVID).

OSNI

Ordnance Survey Northern Ireland.

OSTF

Ordnance Survey Transfer Format. A UK data transfer format previously used by Ordnance Survey for the supply of digital data to customers.

Overlay

A set of graphical data that can be superimposed on to another set of graphical data, eg. user map data on to Ordnance Survey map data. Sometimes used as a synonym for layer, particularly when information is grouped according to colour or some other representational attribute.

Pan

The ability to move the map surface across the screen.

Photogrammetry

The science, art and technology of obtaining reliable measurements and maps from photographs.

Pixel

A picture element - the smallest area of a graphics display that can be addressed.

Point

A zero-dimensional object that specifies a map location through a set of co-ordinates.

Polygon	A closed, two-dimensional figure with three or more sides and intersections; an enclosed geographic area such as a land parcel or political jurisdiction.
Prime contractor	The principal supply-side party to a contract.
Protocol	In data communications, a protocol is a set of rules which determine the formats and conventions by which information may be exchanged between different systems.
Prototyping	In system development, modelling a potential system.
Raster	A format for storing, processing, and displaying graphic data in which graphic images are stored as values for uniform grid cells or pixels.
RDBMS	Relational Database Management System - a DBMS for a relational database.
Relational database	A database organised and accessed according to the relationship between data items. Data is held as rows and columns in tables, and accessed by a series of indexes. By selecting or combining data from different tables, different views of the data can be presented to the user.
RISC	A Reduced Instruction Set Computer is a processor which has a limited instruction set designed such that all instructions can be executed in a single machine cycle.
Rubber sheeting	A process that geometrically adjusts map features to enable a digital map to fit a designated base. The process uses mathematical operations to minimize distortion.
Scanning	A computer graphics technique of generating a display image by a line-by-line sweep across the entire display space; eg the generation of a television screen picture (see raster and digitising).
Spaghetti data	A simple vector data structure comprising feature codes and co-ordinates without any topology.

SPARC	Scalable Processor ARChitecture - a proprietary RISC architecture from Sun Microsystems that has been licensed to other suppliers.
Spatial information	Information which includes a reference to a two or three dimensional position in space as one of its attributes.
Spatial reference	Co-ordinate, textual description or a codified name by which information can be related to a specific position or location on the Earth's surface.
SQL	Structured Query Language is an ISO standard. It is a query language interface for relational databases. SQL is used to define, access and manipulate data stored in these databases.
SSADM	Originally developed by CCTA, the Structured Systems Analysis and Design Method is a non-proprietary structured set of procedural, technical and documentation standards designed specifically for analysing business needs and undertaking software development.
Terrain modelling	See DTM.
Thematic map	A map depicting one or more specific themes; eg. rainfall and population density.
Topological	Describes a geographic data structure in which the inherent spatial connectivity and adjacency relationships of features are explicitly stored and maintained.
Transformation	The ability to change from one co-ordinate system to another.
Vector	A format for processing and displaying graphic data. Vector data comprise strings of co-ordinates representing the true position of features represented by points, lines and areas.
Workstation	A high performance personal computer aimed typically at scientific and technical users. It usually runs a version of the UNIX operating system.

X Window

X Window is a distributed, network-transparent, device independent multi-tasking windowing and graphics system. It was originally developed by MIT in 1984 and has since been widely adopted, particularly by suppliers of UNIX systems. X Window provides the mechanism for drawing windows, it does not define a particular user interface.

Zoom

The ability to change the scale at which the map is displayed across the screen.

Index

V

Value added data 61
Vector:
  co-ordinates 38
  conversion 18, 38
  data 28, 32, 35, 36, 38, 53
  format 12, 13, 28, 36

X

X Window 20

Printed in the United Kingdom for HSMO
Dd297519 11/93 C3 G3397 10170